A Look At
LINEAR MOTION

Rebecca Woodbury, Ph.D., M.Ed.

Gravitas Publications Inc.

A Look At
LINEAR MOTION

Illustrations: Janet Moneymaker

A Look at Linear Motion
ISBN 978-1-950415-23-6

Published by Gravitas Publications Inc.
Imprint: Real Science-4-Kids
www.gravitaspublications.com
www.realscience4kids.com

RS4K

Photo credits: Cover & Title Page: By Qrisio, AdobeStock; Above & P. 3. By master1305, AdobeStock; P. 5. By Brocreative, AdobeStock; P. 7. By JenkoAtaman, AdobeStock; P. 11. By anael_g, AdobeStock; P. 13. By Qrisio, AdobeStock; P. 17. By master1305, AdobeStock; P. 19. NASA

What happens when you
push a hockey puck on ice?

It floats?

NO!

What happens when you throw a baseball straight over the pitcher's mound?

Do you play baseball?

No. The ball is too big.

What happens when you ride a bike straight along a bike path?

When a hockey puck, a
baseball, and a bike move
in a straight line...

...we say they have

linear motion.

The word **linear** means "in a line." Any object moving in a straight line has **linear motion.**

Ants moving in a straight line along an ant trail have linear motion.

I have linear motion.

A snail climbing straight up a wall has linear motion.

Look! A snail carries its own house!

A sprinter running in a straight path has linear motion.

A rocket launching

has linear motion.

Straight up
into space!

What else can you think

of that has linear motion?

Linear motion?

How to say science words

line   (LIYN)

linear   (LIH-nee-uhr)

motion   (MOH-shuhn)

science   (SIY-uhns)

straight   (STRAYT)

www.ingramcontent.com/pod-product-compliance
Lightning Source LLC
Chambersburg PA
CBHW040148200326
41520CB00028B/7531